UNITED STATES IN SPACE: NATIONAL SECURITY STRATEGY AND NATIONAL SPACE POLICY

SPACE SCIENCE, EXPLORATION AND POLICIES

Additional books in this series can be found on Nova's website
under the Series tab.

Additional E-books in this series can be found on Nova's website
under the E-books tab.

DEFENSE, SECURITY AND STRATEGIES

Additional books in this series can be found on Nova's website
under the Series tab.

Additional E-books in this series can be found on Nova's website
under the E-books tab.

SPACE SCIENCE, EXPLORATION AND POLICIES

UNITED STATES IN SPACE: NATIONAL SECURITY STRATEGY AND NATIONAL SPACE POLICY

ANDREW E. OLIVEA
EDITOR

Nova Science Publishers, Inc.
New York

Library of Congress Cataloging-in-Publication Data

United States in space : national security strategy and national space policy / editor, Andrew E. Olivea.
    p. cm.
  Includes bibliographical references and index.
  ISBN 978-1-61324-303-9 (softcover : alk. paper) 1. Astronautics and state--United States. 2. National security--United States. I. Olivea, Andrew E.
  TL789.8.U5U5165 2011
  358'.8--dc23
                            2011012583

Published by Nova Science Publishers, Inc. † New York

CONTENTS

PREFACE

During the past fifty years, U.S. leadership in space activities has benefited the global economy, enhanced our national security, strengthened international relationships, advanced scientific discovery and improved our way of life. Space capabilities provide the U.S. and our allies unprecedented advantages in national decision-making, military operations and homeland security. Space systems provide national security decision-makers with unfettered global access and create a decision advantage by enabling a rapid and tailored response to global challenges. Moreover, space systems are vital to monitoring strategic and military developments as well as supporting treaty monitoring and arms control verification. This book examines the United States national space policy and national security strategy

Chapter 1- During the past 50 years, U.S. leadership in space activities has benefited the global economy, enhanced our national security, strengthened international relationships, advanced scientific discovery, and improved our way of life.

Space capabilities provide the United States and our allies unprecedented advantages in national decision-making, military operations, and homeland security. Space systems provide national security decision-makers with unfettered global access and create a decision advantage by enabling a rapid and tailored response to global challenges. Moreover, space systems are vital to monitoring strategic and military developments as well as supporting treaty monitoring and arms control verification. Space systems are also critical in our ability to respond to natural and man-made disasters and monitor longterm environmental trends. Space systems allow people and governments around the world to see with clarity, communicate with certainty, navigate with accuracy, and operate with assurance.

Chapter 2- WASHINGTON, Feb. 4, 2011 – The National Security Space Strategy released today responds to the realities of a space environment that is increasingly crowded, challenging and competitive, said senior Defense Department officials.

"The National Security Space Strategy represents a significant departure from past practice," Defense Secretary Robert M. Gates said in a DOD news release issued today. "It is a pragmatic approach to maintain the advantages we derive from space while confronting the new challenges we face."

Chapter 3- The space age began as a race for security and prestige between two superpowers . The opportunities were boundless, and the decades that followed have seen a radical transformation in the way we live our daily lives, in large part due to our use of space . Space systems have taken us to other celestial bodies and extended humankind's horizons back in time to the very first moments of the universe and out to the galaxies at its far reaches . Satellites contribute to increased transparency and stability among nations and provide a vital communications path for avoiding potential conflicts . Space systems increase our knowledge in many scientific fields, and life on Earth is far better as a result .

Chapter 4- The majority of large-scale acquisition programs in the Department of Defense's (DOD) space portfolio have experienced problems during the past two decades that have driven up costs by billions of dollars, stretched schedules by years, and increased technical risks. To address the cost increases, DOD altered its acquisitions by reducing the number of satellites it intended to buy, reducing the capabilities of the satellites, or terminating major space systems acquisitions. Moreover, along with the cost increases, many space acquisitions are experiencing significant schedule delays—as much as 8 years—resulting in potential capability gaps in areas such as missile warning, military communications, and weather monitoring. This testimony focuses on

- the status of space acquisitions,
- causal factors of acquisition problems, and
- efforts underway to improve acquisitions.

In: United States in Space   ISBN: 978-1-61324-303-9
Editor: Andrew E. Olivea   © 2011 Nova Science Publishers, Inc.

Chapter 1

NATIONAL SECURITY SPACE STRATEGY: UNCLASSIFIED SUMMARY[*]

United States Department of Defense

During the past 50 years, U.S. leadership in space activities has benefited the global economy, enhanced our national security, strengthened international relationships, advanced scientific discovery, and improved our way of life.

Space capabilities provide the United States and our allies unprecedented advantages in national decision-making, military operations, and homeland security. Space systems provide national security decision-makers with unfettered global access and create a decision advantage by enabling a rapid and tailored response to global challenges. Moreover, space systems are vital to monitoring strategic and military developments as well as supporting treaty monitoring and arms control verification. Space systems are also critical in our ability to respond to natural and man-made disasters and monitor longterm environmental trends. Space systems allow people and governments around the world to see with clarity, communicate with certainty, navigate with accuracy, and operate with assurance.

Maintaining the benefits afforded to the United States by space is central to our national security, but an evolving strategic environment increasingly challenges U.S. space advantages. Space, a domain that no nation owns but on

---

[*] This is an edited, reformatted and augmented version of the United States Department of Defense's publication, dated January 2011.

which all rely, is becoming increasingly congested, contested, and competitive. These challenges, however, also present the United States with opportunities for leadership and partnership. Just as the United States helped promote space security in the 20th century, we will build on this foundation to embrace the opportunities and address the challenges of this century.

The National Security Space Strategy charts a path for the next decade to respond to the current and projected space strategic environment. Leveraging emerging opportunities will strengthen the U.S. national security space posture while maintaining and enhancing the advantages the United States gains from space.

Our strategy requires active U.S. leadership enabled by an approach that updates, balances, and integrates all of the tools of U.S. power. The Department of Defense (DoD) and the Intelligence Community (IC), in coordination with other departments and agencies, will implement this strategy by using it to inform planning, programming, acquisition, operations, and analysis.

Robert M.Gates
Secretary of Defense

James R. Clapper
Director of National Intelligence

THE STRATEGIC ENVIRONMENT

"The now-ubiquitous and interconnected nature of space capabilities and the world's growing dependence on them mean that irresponsible acts in space can have damaging consequences for all of us."

- 2010 National Space Policy

Space is vital to U.S. national security and our ability to understand emerging threats, project power globally, conduct operations, support diplomatic efforts, and enable global economic viability. As more nations and non-state actors recognize these benefits and seek their own space or counterspace capabilities, we are faced with new opportunities and new challenges in the space domain.

The current and future strategic environment is driven by three trends – space is becoming increasingly *congested, contested,* and *competitive.*

Space is increasingly *congested.* Growing global space activity and testing of China's destructive anti-satellite (ASAT) system have increased congestion in important areas in space. DoD tracks approximately 22,000 man-made objects in orbit, of which 1,100 are active satellites (see Figure 1). There may be as many as hundreds of thousands of additional pieces of debris that are too small to track with current sensors. Yet these smaller pieces of debris can damage satellites in orbit.

Today's space environment contrasts with earlier days of the space age in which only a handful of nations needed to be concerned with congestion. Now there are approximately 60 nations and government consortia that own and operate satellites, in addition to numerous commercial and academic satellite operators (see Figure 2). This congestion – along with the effects of operational use, structural failures, accidents involving space systems, and irresponsible testing or employment of debris-producing destructive ASATs – is complicating space operations for all those that seek to benefit from space.

Increased congestion was highlighted by the 2009 collision between a Russian government Cosmos satellite and a U.S. commercial Iridium satellite. The collision created approximately 1,500 new pieces of trackable space debris, adding to the more than 3,000 pieces of debris created by the 2007 Chinese ASAT test. These two events greatly increased the cataloged population of orbital debris.

Another area of increasing congestion is the radiofrequency spectrum. Demand for radiofrequency spectrum to support worldwide satellite services is expected to grow commensurate with the rapid expansion of satellite services and applications. As many as 9,000 satellite communications transponders are expected to be in orbit by 2015. As the demand for bandwidth increases and more transponders are placed in service, the greater the probability of radiofrequency interference and the strain on international processes to minimize that interference.

Space is increasingly *contested* in all orbits. Today space systems and their supporting infrastructure face a range of man-made threats that may deny, degrade, deceive, disrupt, or destroy assets. Potential adversaries are seeking to exploit perceived space vulnerabilities. As more nations and non-state actors develop counterspace capabilities over the next decade, threats to U.S. space systems and challenges to the stability and security of the space environment will increase. Irresponsible acts against space systems could have implications beyond the space domain, disrupting worldwide services upon which the civil and commerical sectors depend.

Space is increasingly *competitive*. Although the United States still maintains an overall edge in space capabilities, the U.S. competitive advantage has decreased as market-entry barriers have lowered (see Figure 3). The U.S. technological lead is eroding in several areas as expertise among other nations increases. International advances in space technology and the associated increase in foreign availability of components have put increased importance on the U.S. export control review process to ensure the competitiveness of the U.S. space industrial base while also addressing national security needs.

U.S. suppliers, especially those in the second and third tiers, are at risk due to inconsistent acquisition and production rates, long development cycles, consolidation of suppliers under first-tier prime contractors, and a more competitive foreign market. A decrease in specialized suppliers further challenges U.S. abilities to maintain assured access to critical technologies, avoid critical dependencies, inspire innovation, and maintain leadership advantages. All of these issues are compounded by challenges in recruiting, developing, and retaining a technical workforce.

STRATEGIC OBJECTIVES

In executing the National Space Policy, our National Security Space Strategy seeks to maintain and enhance the national security benefits we derive from our activities and capabilities in space while addressing and shaping the strategic environment and strengthening the foundations of our enterprise. The U.S. defense and intelligence communities will continue to rely on space systems for military operations, intelligence collection, and related activities; access to these capabilities must be assured. We must address the growing challenges of the congested, contested, and competitive space environment while continuing our leadership in the space domain.

Our strategy is derived from the principles and goals found in the National Space Policy and builds on the strategic approach laid out in the National Security Strategy. Specifically, our national security space objectives are to:

- Strengthen safety, stability, and security in space;
- Maintain and enhance the strategic national security advantages afforded to the United States by space; and
- Energize the space industrial base that supports U.S. national security.

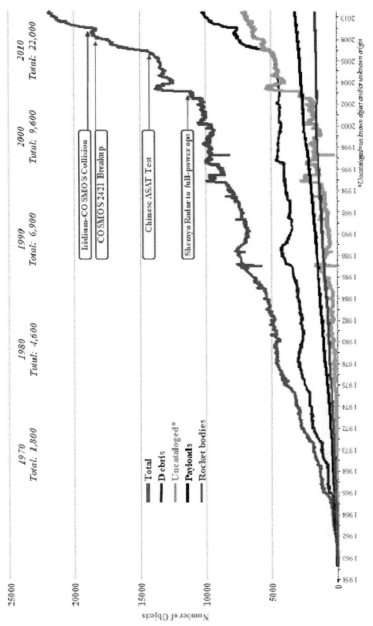

* Uncataloged= unknown object and/or unknown origin.

Source: Joint Space Operations Center.

Figure 1. Satellite Catalog Growth.

Source: National Air and Space Intelligence Center.

Figure 2. Number of Nations and Government Consortia Operating in Space.

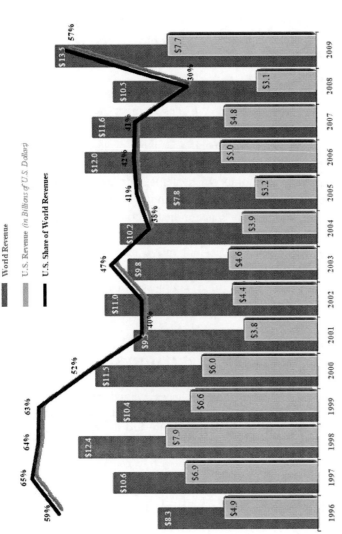

Notes: Revenue figures are in-year estimates, not adjusted for inflation over time. Satellite Manufacturing revenues are recorded in the year the satellite is delivered/launched, not when contract is awarded or interim payments are transacted. World revenue includes U.S. revenue.

Source: Satellite Industry Association.

Figure 3. U.S. versus World Satellite Manufacturing Revenues, 1996–2009.

We seek a safe space environment in which all can operate with minimal risk of accidents, breakups, and purposeful interference. We seek a stable space environment in which nations exercise shared responsibility to act as stewards of the space domain and follow norms of behavior. We seek a secure space environment in which responsible nations have access to space and the benefits of space operations without need to exercise their inherent right of self-defense.

We seek to ensure national security access to space and use of space capabilities in peace, crisis, or conflict. We seek to meet the needs of national leaders and intelligence and military personnel, irrespective of degradation of the space environment or attacks on specific systems or satellites. Enhancing these benefits requires improving the foundational activities of our national security space enterprise – including our systems, our acquisition processes, our industrial base, our technology innovation, and our space professionals.

A resilient, flexible, and healthy space industrial base must underpin all of our space activities. We seek to foster a space industrial base comprised of skilled professionals who deliver those innovative technologies and systems that enable our competitive advantage. Our space system developers, operators, and analysts must deliver, field, and sustain national security space capabilities for the 21st century.

STRATEGIC APPROACHES

"To promote security and stability in space, we will pursue activities consistent with the inherent right of self-defense, deepen cooperation with allies and friends, and work with all nations toward the responsible and peaceful use of space."

- 2010 National Security Strategy

The National Security Space Strategy draws upon all elements of national power and requires active U.S. leadership in space. The United States will pursue a set of interrelated strategic approaches to meet our national security space objectives:

- Promote responsible, peaceful, and safe use of space;
- Provide improved U.S. space capabilities;

- Partner with responsible nations, international organizations, and commercial firms;
- Prevent and deter aggression against space infrastructure that supports U.S. national security; and
- Prepare to defeat attacks and to operate in a degraded environment.

Promoting Responsible, Peaceful, and Safe Use of Space

"All nations have the right to use and explore space, but with this right also comes responsibility. The United States, therefore, calls on all nations to work together to adopt approaches for responsible activity in space to preserve this right for the benefit of future generations."

- 2010 National Space Policy

As directed in the National Space Policy, the United States will promote the responsible, peaceful, and safe use of space as the foundational step to addressing the congested and contested space domain and enabling other aspects of our approach. We will encourage allies, partners, and others to do the same. As more nations, international organizations, and commercial firms field or aspire to field space capabilities, it is increasingly important that they act responsibly, peacefully, and safely in space. At the same time, they must be reassured of U.S. intentions to act likewise. We will encourage responsible behavior in space and lead by the power of our example. Moreover, U.S. diplomatic engagements will enhance our ability to cooperate with our allies and partners and seek common ground among all space-faring nations.

The United States will support development of data standards, best practices, transparency and confidence-building measures, and norms of behavior for responsible space operations. We will consider proposals and concepts for arms control measures if they are equitable, effectively verifiable, and enhance the national security of the United States and its allies. We believe setting pragmatic guidelines for safe activity in space can help avoid collisions and other debris-producing events, reduce radiofrequency interference, and promote security and stability in the space domain – all of which are in the interests of all nations.

Shared awareness of spaceflight activity must improve in order to foster global spaceflight safety and help prevent mishaps, misperceptions, and mistrust. The United States is the leader in space situational awareness (SSA) and can use its knowledge to foster cooperative SSA relationships, support safe space operations, and protect U.S. and allied space capabilities and operations.

DoD will continue to improve the quantity and quality of the SSA information it obtains and expand provision of safety of flight services to U.S. Government agencies, other nations, and commercial firms. DoD will encourage other space operators to share their spaceflight safety data. DoD, in coordination with other government agencies, will seek to establish agreements with other nations and commercial firms to maintain and improve space object databases, pursue common international data standards and data integrity measures, and provide services and disseminate orbital tracking information, including predictions of space object conjunction, to enhance spaceflight safety for all parties.

Providing Improved U.S. Space Capabilities

"Being able to deliver capability cost-effectively when it is needed improves mission effectiveness, provides leadership with flexibility in making investments, and precludes gaps in necessary capabilities."

- 2009 National Intelligence Strategy

U.S. space capabilities will continue to be fundamental for national security. DoD and the IC will identify, improve, and prioritize investments in those capabilities that garner the greatest advantages. We will develop, acquire, field, operate, and sustain space capabilities to deliver timely and accurate space services to a variety of customers, from soldiers to national decision-makers. We will enhance interoperability and compatibility of existing national security systems, across operational domains and mission areas, to maximize efficiency of our national security architecture; we will ensure these characteristics are built into future systems. We will ensure that data collection and products are released at the lowest possible classification to maximize their usefulness to the user community.

Ensuring U.S. capabilities are developed and fielded in a timely, reliable, and responsive manner is critical for national decision-makers to act on time-sensitive and accurate information, for military forces to plan and execute effective operations, and for the IC to enable all of the above with timely indications and warning. Improving our acquisition processes, energizing the U.S. space industrial base, enhancing technological innovation, and deliberately developing space professionals are critical enablers to maintaining U.S. space leadership.

In cooperation with our industrial base partners, DoD and the IC will revalidate current measures and implement new measures, where practicable, to stabilize program acquisition more effectively and improve our space acquisition processes. We will reduce programmatic risk through improved management of requirements. We will use proven best practices of systems engineering, mission assurance, contracting, technology maturation, cost estimating, and financial management to improve system acquisition, reduce the risk of mission failure, and increase successful launch and operation of our space systems.

Mission permitting, we will synchronize the planning, programming, and execution of major acquisition programs with other DoD and IC processes to improve efficiencies and overall performance of our acquisition system and industrial base. DoD and the IC will evaluate the requirements and analysis of alternatives processes to ensure a range of affordable solutions is considered and to identify requirements for possible adjustment. The requirements process must produce combinations of material and non-material solutions. Realistic cost and schedule estimates must inform the President's annual budget request. Human resources processes must provide the right personnel for successful execution.

We seek to foster a U.S. space industrial base that is robust, competitive, flexible, healthy, and delivers reliable space capabilities on time and on budget. DoD and the IC, in concert with the civil space sector, will better manage investments across portfolios to ensure the industrial base can sustain those critical technologies and skills that produce the systems we require. Additionally, we will continue to explore a mix of capabilities with shorter development cycles to minimize delays, cut cost growth, and enable more rapid technology maturation, innovation, and exploitation.

A key aspect of energizing the U.S. space industrial base is to reform U.S. export controls to address technology security and global competitiveness. Export controls have a far-reaching impact on national security interests, as they help deter illicit efforts by others to obtain and use the materials, technology, and know-how that are vital to our national security. Export controls, however, can also affect the health and welfare of the industrial base, in particular second-tier and third-tier suppliers. Reforming export controls will facilitate U.S. firms' ability to compete to become providers-of-choice in the international marketplace for capabilities that are, or will soon become, widely available globally, while strengthening our ability to protect the most significant U.S. technology advantages. In particular, as new opportunities arise for international collaboration, a revised export control system will better enable the domestic firms competing for these contracts. Revised export control policies will address U.S. firms' ability to export space-related items generally available in the global marketplace, consistent with U.S. policy and international commitments.

We will continue to pursue, adapt, and evolve the unique technologies, innovative exploitation techniques, and diverse applications that give the United States its strategic advantage in space. The United States seeks to maintain and enhance access to those global and domestic technologies needed for national security space systems. We will do so by expanding technology partnerships with the academic community, industry, U.S. and partner governments, mission customers, and other centers of technical excellence and innovation, consistent with U.S. policy, technology transfer objectives, and international commitments. To advance the science and technology that enables U.S. space capabilities, we will continue to assess global technology trends to find emerging technologies and potential breakthroughs. We will explore new applications of current technologies and the development of unique, innovative technologies and capabilities. We will improve the transition of scientific research and technology development to the operational user and into major system acquisition. To the extent practicable, we will also facilitate the incorporation of these capabilities and technologies into appropriate domestic space programs.

People are our greatest asset. To support the range of national security space activities, we will develop current and future national security space professionals – our "space cadre" – who can acquire capabilities, operate systems, analyze information, and succeed in a congested, contested, and competitive environment. We will build a more diverse and balanced workforce among military, civilian, and contractor components. These professionals must be educated, experienced, and trained in the best practices of their field – whether it is planning, programming, acquisition, manufacturing, operations, or analysis.

We will continue to encourage students at all levels to pursue technical coursework as a foundation for space-related career fields. Working with other departments and agencies, we will synchronize our science, technology, engineering, and mathematics (STEM) education initiatives with sound education investments to ensure an ample supply of space professionals with appropriate skills and capabilities. We will encourage our space professionals to participate in STEM outreach and mentoring programs.

We will continue to develop structured personnel development programs to expand, track, and sustain our space expertise, employing focused education and training as well as purposeful utilization of our people to offer a broad range of experiential opportunities. We will further professional development by growing, rewarding, and retaining scientific and technical expertise and professional leadership. We will support an entrepreneurial ethos by encouraging initiative, innovation, collaboration, resourcefulness, and resilience. As national security space priorities shift, we will continue to educate and train the workforce to align with new priorities.

Partnering with Responsible Nations, International Organizations, and Commercial Firms

"[E]xplore opportunities to leverage growing international and commercial expertise to enhance U.S. capabilities and reduce the vulnerability of space systems and their supporting ground infrastructure."

- 2010 Quadrennial Defense Review

The evolving strategic environment allows for additional opportunities to partner with responsible nations, international organizations, and commercial firms. DoD and the IC will continue to partner with others to augment the U.S. national security space posture across many mission areas. This includes looking for opportunities to leverage or work in conjunction with partnerships pursued by U.S. Government civil space agencies. By sharing or exchanging capabilities, data, services, personnel, operations, and technology, we can ensure access to information and services from a more diverse set of systems – an advantage in a contested space environment. We will promote appropriate cost-sharing and risk-sharing partnerships to develop and share capabilities. Decisions on partnering will be consistent with U.S. policy and international commitments and consider cost, protection of sources and methods, and effects on the U.S. industrial base.

Partnering with other nations also is essential to ensuring global access to the radiofrequency spectrum and related orbital assignments and promoting the responsible, peaceful, and safe use of outer space. Nations gain international acceptance of their use of the radiofrequency spectrum and satellite orbits through the International Telecommunication Union (ITU). Registering satellite networks with the ITU can help prevent and, if necessary, address radiofrequency interference.

The United States will lead in building coalitions of like-minded space-faring nations and, where appropriate, work with international institutions to do so. With our allies, we will explore the development of combined space doctrine with principles, goals, and objectives that, in particular, endorse and enable the collaborative sharing of space capabilities in crisis and conflict. We will seek to expand mutually beneficial agreements with key partners to utilize existing and planned capabilities that can augment U.S. national security space capabilities. We will pursue increased interoperability, compatibility, and integration of partner nations into appropriate DoD and IC networks to support information sharing and collective endeavors, taking affordability and mutual benefit into account. At the same time, U.S. military and intelligence personnel will ensure the appropriate review and release of classified information to enhance partner access to space information.

We will actively promote the sale of U.S.-developed capabilities to partner nations and the integration of those capabilities into existing U.S. architectures and networks. Posturing our domestic industry to develop these systems will also enable the competitiveness of the U.S. industrial base.

We will explore sharing space-derived information as "global utilities" with partnered nations. As we do today with the positioning, navigation, and timing services of the Global Positioning System, we will provide services derived from selected space systems and enhance those services through partnerships. We will continue to share SSA information to promote responsible and safe space operations. We will also pursue enhanced sharing of other space services such as missile warning and maritime domain awareness. We may seek to establish a collaborative missile warning network to detect attacks against our interests and those of our allies and partners.

Strategic partnerships with commercial firms will continue to enable access to a more diverse, robust, and distributed set of space systems and provide easily releasable data. Strategic partnerships with commercial firms will be pursued in areas that both stabilize costs and improve the resilience of space architectures upon which we rely. Innovative approaches will be explored for their utility in meeting government performance requirements in a cost-effective and timely manner. We will rely on proven commercial capabilities to the maximum extent practicable, and we will modify commercial capabilities to meet government requirements when doing so is more cost-effective and timely for the government. We will develop space systems only when there is no suitable, cost-effective commercial alternative or when national security needs dictate

Preventing and Deterring Aggression against Space Infrastructure that Supports U.S. National Security

> "U.S. forces must be able to deter, defend against, and defeat aggression by potentially hostile nation-states. This capability is fundamental to the nation's ability to protect its interests and to provide security in key regions."
>
> - 2010 Quadrennial Defense Review

Given the degree to which the United States relies on space systems and supporting infrastructure for national security, we must use a multilayered approach to prevent and deter aggression. We seek to enhance our national capability to dissuade and deter the development, testing, and employment of counterspace systems and prevent and deter aggression against space systems and supporting infrastructure that support U.S. national security.

Many elements of this strategy contribute to this approach. We will: support diplomatic efforts to promote norms of responsible behavior in space; pursue international partnerships that encurge potential adversary restraint; improve our ability to attribute attacks; strengthen the resilience of our architectures to deny the benefits of an attack; and retain the right to respond, should deterrence fail.

DoD and the IC will support the diplomatic and public diplomacy efforts of the Department of State to promote the responsible use of space and discourage activities that threaten the safety, stability, and security of the space domain. We will also work with the Department of State and other appropriate U.S. Government agencies to strengthen alliances with other space-faring nations and pursue partnerships with commercial firms and international organizations.

We will improve our intelligence posture – predictive awareness, characterization, warning, and attribution – to better monitor and attribute activities in the space domain. Thus, SSA and foundational intelligence will continue to be top priorities, as they underpin our ability to maintain awareness of natural disturbances and the capabilities, activities, and intentions of others. We will also enable and develop intelligence professionals who can provide greater scope, depth, and quality of intelligence collection and analysis.

We will seek to deny adversaries meaningful benefits of attack by improving cost-effective protection and strengthening the resilience of our architectures. Partnerships with other nations, commercial firms, and international organizations, as well as alternative U.S. Government approaches such as cross-domain solutions, hosted payloads, responsive options, and other innovative solutions, can deliver capability, should our space systems be attacked. This also will enable our ability to operate in a degraded space environment.

Finally, the United States will retain the right and capabilities to respond in self-defense, should deterrence fail. We will use force in a manner that is consistent with longstanding principles of international law, treaties to which the United States is a party, and the inherent right of self defense.

Preparing to Defeat Attacks and Operate in a Degraded Environment

> "Increase assurance and resilience of mission-essential functions enabled by commercial, civil, scientific, and national security spacecraft and supporting infrastructure against disruption, degradation, and destruction, whether from environmental, mechanical, electronic, or hostile causes."
>
> - 2010 National Space Policy

We believe it is in the interests of all space-faring nations to avoid hostilities in space. In spite of this, some actors may still believe counterspace actions could provide military advantage. Our military and intelligence capabilities must be prepared to "fight through" a degraded environment and defeat attacks targeted at our space systems and supporting infrastructure. We must deny and defeat an adversary's ability to achieve its objectives.

As we invest in next generation space capabilities and fill gaps in current capabilities, we will include resilience as a key criterion in evaluating alternative architectures. Resilience can be achieved in a variety of ways, to include cost-effective space system protection, cross-domain solutions, hosting payloads on a mix of platforms in various orbits, drawing on distributed international and commercial partner capabilities, and developing and maturing responsive space capabilities. We will develop the most feasible, mission-effective, and fiscally sound mix of these alternatives.

To make the most effective use of space protection resources, we will identify and prioritize protection for vital space missions supporting national security requirements. We will implement cost-effective protection commensurate with threat, system use, and impact of loss – applied to each segment of our space systems and supporting infrastructure.

To enhance resilience, we will continue to develop mission-effective alternatives, including land, sea, air, space, and cyber-based alternatives for critical capabilities currently delivered primarily through space-based platforms. In addition, we will seek to establish relationships and agreements whereby we can access partner capabilities if U.S. systems are degraded or unavailable. We will be prepared to use these capabilities to ensure the timely continuity of services in a degraded space environment.

Preparing for attacks must extend to the people and processes relying on space information, operating our space systems, and analyzing space-derived information. We will improve the ability of U.S. military and intelligence agencies to operate in a denied or degraded space environment through focused education, training, and exercises and through new doctrine and tactics, techniques, and procedures (TTPs).

IMPLEMENTATION

Consistent with the guidance provided by the President in the National Space Policy, DoD and the IC will implement the National Security Space Strategy by using it to inform future planning, programming, acquisition, operations, and analysis guidance. DoD and the IC will work with other U.S. Government agencies and departments, as well as foreign governments and commercial partners, to update, balance, and integrate all of the tools of U.S. power. We will evolve policies, strategies, and doctrine pertaining to national security space.

Implementation plans will be developed based on feasibility and affordability assessments and cost, benefit, and risk analyses. Further, the impact of plans on manning, operations, and programs will be understood prior to implementation. As stated in the National Security Strategy, our ability to achieve long-term goals for space depends upon our fiscal responsibility and making tough choices, such as between capability and survivability.

CONCLUSION – A NEW TYPE OF LEADERSHIP

"Our national security strategy is, therefore, focused on renewing American leadership so that we can more effectively advance our interests in the 21st century. We will do so by building on the sources of our strength at home, while shaping an international order that can meet the challenges of our time."

- 2010 National Security Strategy

The United States will retain leadership in space by strengthening our posture at home and collaborating with others worldwide. Just as U.S. national security is built upon maintaining strategic advantages, it is also increasingly predicated on active U.S. leadership of alliance and coalition efforts in peacetime, crisis, and conflict.

Active U.S. leadership in space requires a whole-of-government approach that integrates all elements of national power, from technological prowess and industrial capacity to alliance building and diplomatic engagement. Leadership cannot be predicated on declaratory policy alone. It must build upon a willingness to maintain strategic advantages while working with the international community to develop collective norms, share information, and collaborate on capabilities.

U.S. leadership in space can help the United States and our partners address the challenges posed by a space domain that is increasingly congested, contested, and competitive. Our strategy seeks to address this new environment through its set of interrelated approaches:

- We seek to address *congestion* by establishing norms, enhancing space situational awareness, and fostering greater transparency and information sharing. Our words and deeds should reassure our allies and the world at large of our intent to act peacefully and responsibly in space and encourage others to do the same.

- We seek to address the *contested* environment with a multilayered deterrence approach. We will support establishing international norms and transparency and confidence-building measures in space, primarily to promote spaceflight safety but also to dissuade and impose international costs on aggressive behavior. We will improve and protect vital U.S. space capabilities while using interoperability, compatibility, and integration to create coalitions and alliances of responsible space-faring nations. We will improve our capability to attribute attacks and seek to deny meaningful operational benefits from such attacks. We will retain the right and capabilities to respond in self-defense, should deterrence fail.
- We seek to address *competition* by enhancing our own capabilities, improving our acquisition processes, fostering a healthy U.S. industrial base, and strengthening collaboration and cooperation.

Our objectives are to improve safety, stability, and security in space; to maintain and enhance the strategic national security advantages afforded to the United States by space; and to energize the space industrial base that supports U.S. national security. Achieving these objectives will mean not only that our military and intelligence communities can continue to use space for national security purposes, but that a community of nations is working toward creating a sustainable and peaceful space environment to benefit the world for years to come.

In: United States in Space
Editor: Andrew E. Olivea

ISBN: 978-1-61324-303-9
© 2011 Nova Science Publishers, Inc.

Chapter 2

NATIONAL SECURITY SPACE STRATEGY TARGETS SAFETY, STABILITY[*]

Cheryl Pellerin

WASHINGTON, Feb. 4, 2011 – The National Security Space Strategy released today responds to the realities of a space environment that is increasingly crowded, challenging and competitive, said senior Defense Department officials.

"The National Security Space Strategy represents a significant departure from past practice," Defense Secretary Robert M. Gates said in a DOD news release issued today. "It is a pragmatic approach to maintain the advantages we derive from space while confronting the new challenges we face."

Ambassador Gregory L. Schulte, the deputy secretary of defense for space policy, told the Pentagon Channel and American Forces Press Service that this is the first national security space strategy co-signed by the secretary of defense and the director of national intelligence.

"Space has changed in fundamental ways, and that requires us to change our strategy," Schulte said. Gates and Director of National Intelligence James R. Clapper "have signed a document that shows the new directions we need to go," he added.

---

[*] This is an edited, reformatted and augmented version of the American Forces Press Service's publication.

The 10-year strategy concludes the congressionally mandated Space Posture Review by providing strategic objectives and approaches for national security space.

The Defense Department and the intelligence community submitted an interim report to Congress in March that delayed a review of national security space policy and objectives until after the release of the U.S. National Space Policy in June.

Perhaps the strategy's most important message, Schulte said, "is that we have to think differently about how we operate in space."

For example, he said, "we have to think about how to encourage other countries to act responsibly in space and how the United States can provide leadership in that regard.

"Secondly," he added, "we have to think about how we can better leverage the growing amount of foreign commercial capabilities that are now in space. And third, we need to think differently about how to deter others from attacking our space assets."

As in the past, he said, the Defense Department must protect space capabilities to protect the warfighter, whether it's communications, surveillance or global positioning.

"It's space that allows our soldiers to see over the next hill," Schulte said. "It's space that allows us to communicate quickly. It's space that allows us to see whether hostile missiles are launched, so we need to preserve that capability.

"Our goal is to make the peaceful use of space available to all countries," he added, noting that the peaceful use of space includes support for critical defense capabilities.

"Space becomes critical to everything we do, and that's why we're worried that the environment is increasingly challenging," Schulte said. "You have more debris in space and you have countries that are developing counterspace capabilities that can be used against us. That's why this strategy emphasizes the need to protect our capabilities, protect our industrial base and protect the space domain itself."

U.S. Strategic Command officials at Offutt Air Force Base, Neb., are working with other countries and commercial firms to increase situational awareness in space.

"Stratcom was once in charge of delivering nuclear weapons," Schulte said. "Stratcom is now also delivering warnings of potential collisions in space to any variety of countries because we have an interest in preventing more collisions and more debris."

The military also must begin to consider operating in coalitions in space, he said.

"In just about every other domain -- at sea, in the air, on the ground -- we operate with allies and partners. There are good reasons to do it," Schulte said.

Potential partners include members of NATO, whose new 10-year strategic concept issued last year "acknowledged for the first time that access to space is something you can't take for granted," he said.

The Joint Space Operations Center at Vandenberg Air Force Base, Calif., is a focal point for the operational use of worldwide U.S. space forces, Schulte said, and it allows the commander of Stratcom's joint functional component command for space to integrate space power into global military operations.

"We would like to make that into a [combined center]," Schulte said, "where we bring in our closest allies and eventually others, so that like in other domains, we can conduct combined operations."

The 10-year National Security Space Strategy will require at least that long to implement, he said.

"You will see some early indications of it in the president's budget for 2012, and you will see more in his budget for 2013, but ultimately what we're trying to do is affect programs of the services, particularly the Air Force, over the longer term," Schulte said. "We're trying to affect how we train, we're trying to affect how we plan, and we're trying to affect the diplomacy we conduct with the Department of State. So I think you'll see [the strategy] roll out in many different ways. In fact, you're already seeing elements of it."

On Jan. 6, Gates announced that he would use some of the efficiency savings Air Force officials identified to invest in the U.S. launch capability to help in protecting the industrial base, Schulte said.

Defense Department officials are working Australia on sharing of space situational awareness and are talking to the commercial sector about how DOD can host payloads on their satellites, he said. "And we're looking for a whole range of activities to implement the new strategy in a budget-constrained environment," he added

Schulte said to get DOD organized for space, Deputy Defense Secretary William J. Lynn III created the Space Defense Council, to be chaired by Air Force Secretary Michael B. Donley.

"The secretary and the deputy have entrusted to Secretary Donley the role of moving forward with our strategy," he said, "and the Defense Space Council provides a forum to do that."

In: United States in Space
Editor: Andrew E. Olivea

ISBN: 978-1-61324-303-9
© 2011 Nova Science Publishers, Inc.

Chapter 3

NATIONAL SPACE POLICY
OF THE UNITED STATES OF AMERICA

INTRODUCTION

"More than by any other imaginative concept, the mind of man is aroused by the thought of exploring the mysteries of outer space. Through such exploration, man hopes to broaden his horizons, add to his knowledge, improve his way of living on earth."

— President Dwight Eisenhower, June 20, 1958

"Fifty years after the creation of NASA, our goal is no longer just a destination to reach. Our goal is the capacity for people to work and learn and operate and live safely beyond the Earth for extended periods of time, ultimately in ways that are more sustainable and even indefinite. And in fulfilling this task, we will not only extend humanity's reach in space— we will strengthen America's leadership here on Earth."

— President Barack Obama, April 15, 2010

The space age began as a race for security and prestige between two superpowers . The opportunities were boundless, and the decades that followed have seen a radical transformation in the way we live our daily lives, in large part due to our use of space . Space systems have taken us to other celestial bodies and extended humankind's horizons back in time to the very first moments of the universe and out to the galaxies at its far reaches . Satellites contribute to increased transparency and stability among nations and provide a vital communications path for avoiding potential conflicts . Space systems increase our knowledge in many scientific fields, and life on Earth is far better as a result .

The utilization of space has created new markets; helped save lives by warning us of natural disasters, expediting search and rescue operations, and making recovery efforts faster and more effective; made agriculture and natural resource management more efficient and sustainable; expanded our frontiers; and provided global access to advanced medicine, weather forecasting, geospatial information, financial operations, broadband and other communications, and scores of other activities worldwide . Space systems allow people and governments around the world to see with clarity, communicate with certainty, navigate with accuracy, and operate with assurance .

The legacy of success in space and its transformation also presents new challenges . When the space age began, the opportunities to use space were limited to only a few nations, and there were limited consequences for irresponsible or unintentional behavior . Now, we find ourselves in a world where the benefits of space permeate almost every facet of our lives . The growth and evolution of the global economy has ushered in an ever-increasing number of nations and organizations using space . The now-ubiquitous and interconnected nature of space capabilities and the world's growing dependence on them mean that irresponsible acts in space can have damaging consequences for all of us . For example, decades of space activity have littered Earth's orbit with debris; and as the world's space-faring nations continue to increase activities in space, the chance for a collision increases correspondingly .

As the leading space-faring nation, the United States is committed to addressing these challenges . But this cannot be the responsibility of the United States alone . All nations have the right to use and explore space, but with this right also comes responsibility . The United States, therefore, calls on all nations to work together to adopt approaches for responsible activity in space to preserve this right for the benefit of future generations .

From the outset of humanity's ascent into space, this Nation declared its commitment to enhance the welfare of humankind by cooperating with others to maintain the freedom of space .

The United States hereby renews its pledge of cooperation in the belief that with strengthened international collaboration and reinvigorated U .S . leadership, all nations and peoples—space-faring and space-benefiting—will find their horizons broadened, their knowledge enhanced, and their lives greatly improved .

PRINCIPLES

In this spirit of cooperation, the United States will adhere to, and proposes that other nations recognize and adhere to, the following principles:

- It is the shared interest of all nations to act responsibly in space to help prevent mishaps, misperceptions, and mistrust . The United States considers the sustainability, stability, and free access to, and use of, space vital to its national interests . Space operations should be conducted in ways that emphasize openness and transparency to improve public awareness of the activities of government, and enable others to share in the benefits provided by the use of space .
- A robust and competitive commercial space sector is vital to continued progress in space . The United States is committed to encouraging and facilitating the growth of a U .S . commercial space sector that supports U .S . needs, is globally competitive, and advances U .S . leadership in the generation of new markets and innovation-driven entrepreneurship .
- All nations have the right to explore and use space for peaceful purposes, and for the benefit of all humanity, in accordance with international law . Consistent with this principle, "peaceful purposes" allows for space to be used for national and homeland security activities .
- As established in international law, there shall be no national claims of sovereignty over outer space or any celestial bodies . The United States considers the space systems of all nations to have the rights of passage through, and conduct of operations in, space without interference . Purposeful interference with space systems, including supporting infrastructure, will be considered an infringement of a nation's rights .
- The United States will employ a variety of measures to help assure the use of space for all responsible parties, and, consistent with the inherent right of self-defense, deter others from interference and attack, defend our space systems and contribute to the defense of

allied space systems, and, if deterrence fails, defeat efforts to attack them.

GOALS

Consistent with these principles, the United States will pursue the following goals in its national space programs:

- **Energize competitive domestic industries** to participate in global markets and advance the development of: satellite manufacturing; satellite-based services; space launch; terrestrial applications; and increased entrepreneurship .
- **Expand international cooperation** on mutually beneficial space activities to: broaden and extend the benefits of space; further the peaceful use of space; and enhance collection and partnership in sharing of space-derived information .
- **Strengthen stability in space** through: domestic and international measures to promote safe and responsible operations in space; improved information collection and sharing for space object collision avoidance; protection of critical space systems and supporting infrastructures, with special attention to the critical interdependence of space and information systems; and strengthening measures to mitigate orbital debris .
- **Increase assurance and resilience of mission-essential functions** enabled by commercial, civil, scientific, and national security spacecraft and supporting infrastructure against disruption, degradation, and destruction, whether from environmental, mechanical, electronic, or hostile causes .
- **Pursue human and robotic initiatives** to develop innovative technologies, foster new industries, strengthen international partnerships, inspire our Nation and the world, increase humanity's understanding of the Earth, enhance scientific discovery, and explore our solar system and the universe beyond .
- **Improve space-based Earth and solar observation** capabilities needed to conduct science, forecast terrestrial and near-Earth space weather, monitor climate and global change, manage natural resources, and support disaster response and recovery.

All actions undertaken by departments and agencies in implementing this directive shall be within the overall resource and policy guidance provided by the President; consistent with U .S . law and regulations, treaties and other agreements to which the United States is a party, other applicable international law, U .S . national and homeland security requirements, U .S . foreign policy, and national interests; and in accordance with the Presidential Memorandum on Transparency and Open Government .

INTERSECTOR GUIDELINES

In pursuit of this directive's goals, all departments and agencies shall execute the following guidance:

Foundational Activities and Capabilities

- **Strengthen U.S. Leadership In Space-Related Science, Technology, and Industrial Bases** . Departments and agencies shall: conduct basic and applied research that increases capabilities and decreases costs, where this research is best supported by the government; encourage an innovative and entrepreneurial commercial space sector; and help ensure the availability of space-related industrial capabilities in support of critical government functions .
- **Enhance Capabilities for Assured Access To Space.** United States access to space depends in the first instance on launch capabilities . United States Government payloads shall be launched on vehicles manufactured in the United States unless exempted by the National Security Advisor and the Assistant to the President for Science and Technology and Director of the Office of Science and Technology Policy, consistent with established interagency standards and coordination guidelines . Where applicable to their responsibilities departments and agencies shall:
  - o Work jointly to acquire space launch services and hosted payload arrangements that are reliable, responsive to United States Government needs, and cost-effective;
  - o Enhance operational efficiency, increase capacity, and reduce launch costs by investing in the modernization of space launch infrastructure; and

o   Develop launch systems and technologies necessary to assure and sustain future reliable and efficient access to space, in cooperation with U .S . industry, when sufficient U .S . commercial capabilities and services do not exist .

- **Maintain and Enhance Space-based Positioning, Navigation, and Timing Systems.** The United States must maintain its leadership in the service, provision, and use of global navigation satellite systems (GNSS) . To this end, the United States shall:

  o   Provide continuous worldwide access, for peaceful civil uses, to the Global Positioning System (GPS) and its government-provided augmentations, free of direct user charges;

  o   Engage with foreign GNSS providers to encourage compatibility and interoperability, promote transparency in civil service provision, and enable market access for U .S . industry;

  o   Operate and maintain the GPS constellation to satisfy civil and national security needs, consistent with published performance standards and interface specifications . Foreign positioning, navigation, and timing (PNT) services may be used to augment and strengthen the resiliency of GPS; and

  o   Invest in domestic capabilities and support international activities to detect, mitigate, and increase resiliency to harmful interference to GPS, and identify and implement, as necessary and appropriate, redundant and back-up systems or approaches for critical infrastructure, key resources, and mission-essential functions .

- **Develop and Retain Space Professionals.** The primary goals of space professional development and retention are: achieving mission success in space operations and acquisition; stimulating innovation to improve commercial, civil, and national security space capabilities; and advancing science, exploration, and discovery . Toward these ends, departments and agencies, in cooperation with industry and academia, shall establish standards, seek to create opportunities for the current space workforce, and implement measures to develop, maintain, and retain skilled space professionals, including engineering and scientific personnel and experienced space system developers and operators, in government and commercial workforces . Departments and agencies also shall promote and expand public-private partnerships to foster educational achievement in Science,

Technology, Engineering, and Mathematics (STEM) programs, supported by targeted investments in such initiatives .

- Improve Space System Development and Procurement . Departments and agencies shall:

  o Improve timely acquisition and deployment of space systems through enhancements in estimating costs, technological risk and maturity, and industrial base capabilities;
  o Reduce programmatic risk through improved management of requirements and by taking advantage of cost-effective opportunities to test high-risk components, payloads, and technologies in space or relevant environments;
  o Embrace innovation to cultivate and sustain an entrepreneurial U .S . research and development environment; and
  o Engage with industrial partners to improve processes and effectively manage the supply chains .

- **Strengthen Interagency Partnerships.** Departments and agencies shall improve their partnerships through cooperation, collaboration, information sharing, and/or alignment of common pursuits . Departments and agencies shall make their capabilities and expertise available to each other to strengthen our ability to achieve national goals, identify desired outcomes, leverage U .S . capabilities, and develop implementation and response strategies .

International Cooperation

Strengthen U.S. Space Leadership
Departments and agencies, in coordination with the Secretary of State, shall:

- Demonstrate U .S . leadership in space-related fora and activities to: reassure allies of U .S . commitments to collective self-defense; identify areas of mutual interest and benefit; and promote U .S . commercial space regulations and encourage interoperability with these regulations;
- Lead in the enhancement of security, stability, and responsible behavior in space;

- Facilitate new market opportunities for U .S . commercial space capabilities and services, including commercially viable terrestrial applications that rely on government-provided space systems;
- Promote the adoption of policies internationally that facilitate full, open, and timely access to government environmental data;
- Promote appropriate cost- and risk-sharing among participating nations in international partnerships; and
- Augment U .S . capabilities by leveraging existing and planned space capabilities of allies and space partners .

Identify Areas for Potential International Cooperation

Departments and agencies shall identify potential areas for international cooperation that may include, but are not limited to: space science; space exploration, including human space flight activities; space nuclear power to support space science and exploration; space transportation; space surveillance for debris monitoring and awareness; missile warning; Earth science and observation; environmental monitoring; satellite communications; GNSS; geospatial information products and services; disaster mitigation and relief; search and rescue; use of space for maritime domain awareness; and long-term preservation of the space environment for human activity and use .

The Secretary of State, after consultation with the heads of appropriate departments and agencies, shall carry out diplomatic and public diplomacy efforts to strengthen understanding of, and support for, U .S . national space policies and programs and to encourage the foreign use of U .S . space capabilities, systems, and services .

Develop Transparency and Confidence-Building Measures

The United States will pursue bilateral and multilateral transparency and confidence-building measures to encourage responsible actions in, and the peaceful use of, space . The United States will consider proposals and concepts for arms control measures if they are equitable, effectively verifiable, and enhance the national security of the United States and its allies .

Preserving the Space Environment and the Responsible Use of Space

Preserve the Space Environment

For the purposes of minimizing debris and preserving the space environment for the responsible, peaceful, and safe use of all users, the United States shall:

- Lead the continued development and adoption of international and industry standards and policies to minimize debris, such as the United Nations Space Debris Mitigation Guidelines;
- Develop, maintain, and use space situational awareness (SSA) information from commercial, civil, and national security sources to detect, identify, and attribute actions in space that are contrary to responsible use and the long-term sustainability of the space environment;
- Continue to follow the United States Government Orbital Debris Mitigation Standard Practices, consistent with mission requirements and cost effectiveness, in the procurement and operation of spacecraft, launch services, and the conduct of tests and experiments in space;
- Pursue research and development of technologies and techniques, through the Administrator of the National Aeronautics and Space Administration (NASA) and the Secretary of Defense, to mitigate and remove on-orbit debris, reduce hazards, and increase understanding of the current and future debris environment; and
- Require the head of the sponsoring department or agency to approve exceptions to the United States Government Orbital Debris Mitigation Standard Practices and notify the Secretary of State .

Foster the Development of Space Collision Warning Measures

The Secretary of Defense, in consultation with the Director of National Intelligence, the Administrator of NASA, and other departments and agencies, may collaborate with industry and foreign nations to: maintain and improve space object databases; pursue common international data standards and data integrity measures; and provide services and disseminate orbital tracking information to commercial and international entities, including predictions of space object conjunction .

Effective Export Policies

Consistent with the U .S . export control review, departments and agencies should seek to enhance the competitiveness of the U .S . space industrial base while also addressing national security needs .

The United States will work to stem the flow of advanced space technology to unauthorized parties . Departments and agencies are responsible for protecting against adverse technology transfer in the conduct of their programs .

The United States Government will consider the issuance of licenses for space-related exports on a case-by-case basis, pursuant to, and in accordance with, the International Traffic in Arms Regulations, the Export Administration Regulations, and other applicable laws, treaties, and regulations . Consistent with the foregoing space-related items that are determined to be generally available in the global marketplace shall be considered favorably with a view that such exports are usually in the national interests of the United States .

Sensitive or advanced spacecraft-related exports may require a government-to-government agreement or other acceptable arrangement .

Space Nuclear Power

The United States shall develop and use space nuclear power systems where such systems safely enable or significantly enhance space exploration or operational capabilities .

Approval by the President or his designee shall be required to launch and use United States Government spacecraft utilizing nuclear power systems either with a potential for criticality or above a minimum threshold of radioactivity, in accordance with the existing interagency review process . To inform this decision, the Secretary of Energy shall conduct a nuclear safety analysis for evaluation by an ad hoc Interagency Nuclear Safety Review Panel that will evaluate the risks associated with launch and in-space operations .

The Secretary of Energy shall:

- Assist the Secretary of Transportation in the licensing of space transportation activities involving spacecraft with nuclear power systems;
- Provide nuclear safety monitoring to ensure that operations in space are consistent with any safety evaluations performed; and

- Maintain the capability and infrastructure to develop and furnish nuclear power systems for use in United States Government space systems .

Radiofrequency Spectrum and Interference Protection

The United States Government shall:

- Seek to protect U .S . global access to, and operation in, the radiofrequency spectrum and related orbital assignments required to support the use of space by the United States Government, its allies, and U .S . commercial users;
- Explicitly address requirements for radiofrequency spectrum and orbital assignments prior to approving acquisition of space capabilities;
- Seek to ensure the necessary national and international regulatory frameworks will remain in place over the lifetime of the system;
- Identify impacts to government space systems prior to reallocating spectrum for commercial, federal, or shared use;
- Enhance capabilities and techniques, in cooperation with civil, commercial, and foreign partners, to identify, locate, and attribute sources of radio frequency interference, and take necessary measures to sustain the radiofrequency environment in which critical U .S . space systems operate; and
- Seek appropriate regulatory approval under U .S . domestic regulations for United States Government earth stations operating with commercially owned satellites, consistent with the regulatory approval granted to analogous commercial earth stations.

Assurance and Resilience of Mission-Essential Functions

The United States shall:

- Assure space-enabled mission-essential functions by developing the techniques, measures, relationships, and capabilities necessary to maintain continuity of services;
  - o Such efforts may include enhancing the protection and resilience of selected spacecraft and supporting infrastructure;

- Develop and exercise capabilities and plans for operating in and through a degraded, disrupted, or denied space environment for the purposes of maintaining mission-essential functions; and
- Address mission assurance requirements and space system resilience in the acquisition of future space capabilities and supporting infrastructure .

SECTOR GUIDELINES

United States space activities are conducted in three distinct but interdependent sectors: commercial, civil, and national security .

Commercial Space Guidelines

The term "commercial," for the purposes of this policy, refers to space goods, services, or activities provided by private sector enterprises that bear a reasonable portion of the investment risk and responsibility for the activity, operate in accordance with typical market-based incentives for controlling cost and optimizing return on investment, and have the legal capacity to offer these goods or services to existing or potential nongovernmental customers . To promote a robust domestic commercial space industry, departments and agencies shall:

- Purchase and use commercial space capabilities and services to the maximum practical extent when such capabilities and services are available in the marketplace and meet United States Government requirements;
- Modify commercial space capabilities and services to meet government requirements when existing commercial capabilities and services do not fully meet these requirements and the potential modification represents a more cost-effective and timely acquisition approach for the government;
- Actively explore the use of inventive, nontraditional arrangements for acquiring commercial space goods and services to meet United States Government requirements, including measures such as public-private partnerships, hosting government capabilities on commercial

spacecraft, and purchasing scientific or operational data products from commercial satellite operators in support of government missions;

- Develop governmental space systems only when it is in the national interest and there is no suitable, cost-effective U .S . commercial or, as appropriate, foreign commercial service or system that is or will be available;

- Refrain from conducting United States Government space activities that preclude, discourage, or compete with U .S . commercial space activities, unless required by national security or public safety;

- Pursue potential opportunities for transferring routine, operational space functions to the commercial space sector where beneficial and cost-effective, except where the government has legal, security, or safety needs that would preclude commercialization;

- Cultivate increased technological innovation and entrepreneurship in the commercial space sector through the use of incentives such as prizes and competitions;

- Ensure that United States Government space technology and infrastructure are made available for commercial use on a reimbursable, noninterference, and equitable basis to the maximum practical extent;

- Minimize, as much as possible, the regulatory burden for commercial space activities and ensure

- that the regulatory environment for licensing space activities is timely and responsive;

- Foster fair and open global trade and commerce through the promotion of suitable standards and regulations that have been developed with input from U .S . industry;

- Encourage the purchase and use of U .S . commercial space services and capabilities in international cooperative arrangements; and

- Actively promote the export of U .S . commercially developed and available space goods and services, including those developed by small- and medium-sized enterprises, for use in foreign markets, consistent with U .S . technology transfer and nonproliferation objectives .

The United States Trade Representative (USTR) has the primary responsibility in the Federal Government for international trade agreements to which the United States is a party . USTR, in consultation with other relevant departments and agencies, will lead any efforts relating to the negotiation and implementation of trade disciplines governing trade in goods and services related to space .

Civil Space Guidelines

Space Science, Exploration, and Discovery

The Administrator of NASA shall:

- Set far-reaching exploration milestones . By 2025, begin crewed missions beyond the moon, including sending humans to an asteroid . By the mid-2030s, send humans to orbit Mars and return them safely to Earth;
- Continue the operation of the International Space Station (ISS), in cooperation with its international partners, likely to 2020 or beyond, and expand efforts to: utilize the ISS for scientific, technological, commercial, diplomatic, and educational purposes; support activities requiring the unique attributes of humans in space; serve as a continuous human presence in Earth orbit; and support future objectives in human space exploration;
- Seek partnerships with the private sector to enable safe, reliable, and cost-effective commercial spaceflight capabilities and services for the transport of crew and cargo to and from the ISS;
- Implement a new space technology development and test program, working with industry, academia, and international partners to build, fly, and test several key technologies that can increase the capabilities, decrease the costs, and expand the opportunities for future space activities;
- Conduct research and development in support of next-generation launch systems, including new U .S . rocket engine technologies;
- Maintain a sustained robotic presence in the solar system to: conduct scientific investigations of other planetary bodies; demonstrate new technologies; and scout locations for future human missions;
- Continue a strong program of space science for observations, research, and analysis of our Sun, solar system, and universe to enhance knowledge of the cosmos, further our understanding of fundamental natural and physical sciences, understand the conditions

that may support the development of life, and search for planetary bodies and Earth-like planets in orbit around other stars; and

- Pursue capabilities, in cooperation with other departments, agencies, and commercial partners, to detect, track, catalog, and characterize near-Earth objects to reduce the risk of harm to humans from an unexpected impact on our planet and to identify potentially resource-rich planetary objects .

Environmental Earth Observation and Weather

To continue and improve a broad array of programs of space-based observation, research, and analysis of the Earth's land, oceans, and atmosphere:

- The NASA Administrator, in coordination with other appropriate departments and agencies, shall conduct a program to enhance U .S . global climate change research and sustained monitoring capabilities, advance research into and scientific knowledge of the Earth by accelerating the development of new Earth observing satellites, and develop and test capabilities for use by other civil departments and agencies for operational purposes .
- The Secretary of Commerce, through the National Oceanic and Atmospheric Administration (NOAA) Administrator, and in coordination with the NASA Administrator and other appropriate departments and agencies, shall, in support of operational requirements:

  o Transition mature research and development Earth observation satellites to long-term operations;
  o Use international partnerships to help sustain and enhance weather, climate, ocean, and coastal observation from space; and
  o Be responsible for the requirements, funding, acquisition, and operation of civil operational environmental satellites in support of weather forecasting, climate monitoring, ocean and coastal observations, and space weather forecasting . NOAA will primarily utilize NASA as the acquisition agent for operational environmental satellites for these activities and programs .

- The Secretary of Commerce, through the NOAA Administrator, the Secretary of Defense, through the Secretary of the Air Force, and the NASA Administrator shall work together and with their international partners to ensure uninterrupted, operational polar-orbiting environmental satellite observations . The Secretary of Defense shall be responsible for the morning orbit, and the Secretary of Commerce shall be responsible for the afternoon orbit . The departments shall continue to partner in developing and fielding a shared ground system, with the coordinated programs operated by NOAA . Further, the departments shall ensure the continued full sharing of data from all systems .

Land Remote Sensing

The Secretary of the Interior, through the Director of the United States Geological Survey (USGS), shall:

- Conduct research on natural and human-induced changes to Earth's land, land cover, and inland surface waters, and manage a global land surface data national archive and its distribution;
- Determine the operational requirements for collection, processing, archiving, and distribution of land surface data to the United States Government and other users; and
- Be responsible, in coordination with the Secretary of Defense, the Secretary of Homeland Security, and the Director of National Intelligence, for providing remote sensing information related to the environment and disasters that is acquired from national security space systems to other civil government agencies .

In support of these critical needs, the Secretary of the Interior, through the Director of the USGS, and the NASA Administrator shall work together in maintaining a program for operational land remote sensing observations .

The NASA and NOAA Administrators and the Director of the USGS shall:

- Ensure that civil space acquisition processes and capabilities are not unnecessarily duplicated; and

- Continue to develop civil applications and information tools based on data collected by Earth observation satellites . These civil capabilities will be developed, to the greatest extent possible, using known standards and open protocols, and the applications will be made available to the public .

The Secretary of Commerce, through the Administrator of NOAA, shall provide for the regulation and licensing of the operation of commercial sector remote sensing systems .

National Security Space Guidelines

The Secretary of Defense and the Director of National Intelligence, in consultation with other appropriate heads of departments and agencies, shall:

- Develop, acquire, and operate space systems and supporting information systems and networks to support U .S . national security and enable defense and intelligence operations during times of peace, crisis, and conflict;
- Ensure cost-effective survivability of space capabilities, including supporting information systems and networks, commensurate with their planned use, the consequences of lost or degraded capability, the threat, and the availability of other means to perform the mission;
- Reinvigorate U .S . leadership by promoting technology development, improving industrial capacity, and maintaining a robust supplier base necessary to support our most critical national security interests;
- Develop and implement plans, procedures, techniques, and capabilities necessary to assure critical national security space-enabled missions . Options for mission assurance may include rapid restoration of space assets and leveraging allied, foreign, and/or commercial space and nonspace capabilities to help perform the mission;
- Maintain and integrate space surveillance, intelligence, and other information to develop accurate and timely SSA . SSA information shall be used to support national and homeland security, civil space agencies, particularly human space flight activities, and commercial and foreign space operations;

- Improve, develop, and demonstrate, in cooperation with relevant departments and agencies and commercial and foreign entities, the ability to rapidly detect, warn, characterize, and attribute natural and man-made disturbances to space systems of U .S . interest; and
- Develop and apply advanced technologies and capabilities that respond to changes to the threat environment .

The Secretary of Defense shall:

- Be responsible, with support from the Director of National Intelligence, for the development, acquisition, operation, maintenance, and modernization of SSA capabilities;
- Develop capabilities, plans, and options to deter, defend against, and, if necessary, defeat efforts to interfere with or attack U .S . or allied space systems;
- Maintain the capabilities to execute the space support, force enhancement, space control, and force application missions; and
- Provide, as launch agent for both the defense and intelligence sectors, reliable, affordable, and timely space access for national security purposes .

The Director of National Intelligence shall:

- Enhance foundational intelligence collection and single- and all-source intelligence analysis;
- Develop, obtain, and operate space capabilities to support strategic goals, intelligence priorities, and assigned tasks;
- Provide robust, timely, and effective collection, processing, analysis, and dissemination of information on foreign space and supporting information system activities;
- Develop and enhance innovative analytic tools and techniques to use and share information from traditional and nontraditional sources for understanding foreign space-related activities;
- Identify and characterize current and future threats to U .S . space missions for the purposes of enabling effective protection, deterrence, and defense;
- Integrate all-source intelligence of foreign space capabilities and intentions with space surveillance information to produce enhanced intelligence products that support SSA;

- Support national defense and homeland security planning and satisfy operational requirements as a major intelligence mission;
- Support monitoring, compliance, and verification for transparency and confidence-building measures and, if applicable, arms control agreements; and
- Coordinate on any radiofrequency surveys from space conducted by United States Government departments or agencies and review, as appropriate, any radiofrequency surveys from space conducted by licensed private sector operators or by state and local governments .

In: United States in Space   ISBN: 978-1-61324-303-9
Editor: Andrew E. Olivea   © 2011 Nova Science Publishers, Inc.

Chapter 4

SPACE ACQUISITIONS: DOD POISED TO ENHANCE SPACE CAPABILITIES, BUT PERSISTENT CHALLENGES REMAIN IN DEVELOPING SPACE SYSTEMS[*]

United States Government Accountability Office

WHY GAO DID THIS STUDY

The majority of large-scale acquisition programs in the Department of Defense's (DOD) space portfolio have experienced problems during the past two decades that have driven up costs by billions of dollars, stretched schedules by years, and increased technical risks. To address the cost increases, DOD altered its acquisitions by reducing the number of satellites it intended to buy, reducing the capabilities of the satellites, or terminating major space systems acquisitions. Moreover, along with the cost increases, many space acquisitions are experiencing significant schedule delays—as much as 8 years—resulting in potential capability gaps in areas such as missile warning, military communications, and weather monitoring. This testimony focuses on

---

[*] This is an edited, reformatted and augmented version of the United States Government Accountability Office's publication, dated March 10, 2010. These remarks were delievered as testimony given on March 10, 2010. Cristina T. Chaplain, Director Acquisition and Sourcing Management.

- the status of space acquisitions,
- causal factors of acquisition problems, and
- efforts underway to improve acquisitions.

In preparing this testimony, GAO relied on its body of work, including GAO reports on best practices, assessments of individual space programs, common problems affecting space system acquisitions, and the DOD's acquisition policies. We have made numerous recommendations to the DOD in the past on matters relating to overall best practices as well as on individual space program acquisitions. DOD often concurred with our findings and recommendations and has efforts underway to adopt best practices.

WHAT GAO FOUND

A long-standing problem in DOD space acquisitions is that program and unit costs tend to go up significantly from initial cost estimates, while in some cases, the capability that was to be produced declines. This problem persists. However, DOD has made progress on several of its high-risk space programs and is expecting to launch new generations of satellites across various missions over the next 12 months that should significantly advance some capabilities, particularly protected communications and space surveillance. While DOD is having success in readying some satellites for launch, other space acquisition programs currently in development face challenges that could further increase costs and delay targeted delivery dates. Another risk facing DOD space programs over the next few years is the potential for launch delays because of changes being made in the launch sector and an increase in the demand for certain DOD launch vehicles.

Our past work has identified a number of causes for the cost growth and related problems, but several consistently stand out. First, on a broad scale, DOD starts more weapon programs than it can afford, creating a competition for funding that encourages low cost estimating, optimistic scheduling, overpromising, suppressing bad news, and, for space programs, forsaking the opportunity to identify and assess potentially more executable alternatives. Second, DOD has tended to start its space programs too early, that is, before it has the assurance that the capabilities it is pursuing can be achieved within available resources and time constraints. This tendency is caused largely by the funding process, since acquisition programs attract more dollars than efforts concentrating solely on proving technologies. Third, programs have historically attempted to satisfy all requirements in a single step, regardless of the design challenge or the maturity of the technologies necessary to achieve the full capability.

DOD has been working to ensure that its space programs are more executable and produce a better return on investment. Some actions DOD and others have adopted or are pursuing include: the Acquisition Improvement Plan, which lists five initiatives for improving how the Air Force obtains new capabilities; changes in cost estimating that are in line with earlier GAO recommendations; and the Weapon Systems Acquisition Reform Act, which was signed into law in May 2009. However, there are still significant changes to processes, policies, and support needed to ensure reforms can take hold. Recent studies and reviews that have examined national security space have all found that diffuse leadership has a direct impact on the space acquisition process, primarily because it makes it difficult to hold any one person or organization accountable, and there is no single authority to resolve conflicts among the many organizations involved in space programs. Moreover, DOD continues to face gaps in critical technical and program expertise for space. Until both issues are resolved, commitment to reforms may not be sustainable.

Mr. Chairman and Members of the Subcommittee:

I am pleased to be here today to discuss the Department of Defense's (DOD) space acquisitions. Each year, DOD spends billions of dollars to acquire space-based capabilities to support current military and other government operations, as well as to enable DOD to transform the way it collects and disseminates information. Despite the significant investment in space, the majority of large-scale acquisition programs in DOD's space portfolio have experienced problems during the past two decades that have driven up costs by hundreds of millions and even billions of dollars and stretched schedules by years and increased technical risks. To address the cost increases, DOD altered its acquisitions by reducing the number of satellites it intended to buy, reducing the capabilities of the satellites, or terminating major space systems acquisitions. Moreover, along with the cost increases, many space acquisitions have experienced significant schedule delays—of as much as 8 years—resulting in potential capability gaps in areas such as missile warning, military communications, and weather monitoring. These problems persist.

My testimony today will focus on: (1) the status of space acquisitions, (2) the efforts DOD is taking to address causes of problems and increase credibility and success in its space systems acquisitions, and (3) what remains to be done. Notably, DOD has taken the important step of acknowledging the acquisition problems of the past and is taking action to address them, including better management of the acquisition process and oversight of its contractors. Moreover, several high-risk space programs have finally resolved technical and other obstacles and are close to begin delivering capability. However, other space acquisition programs continue to face challenges in meeting their cost and schedule targets and aligning the delivery of space assets with the ground and user systems needed to support and take advantage of new capability. Additionally, it may take years for acquisition improvements to take root and produce benefits that will enable DOD to realize a better return on its investment in space. Lastly, DOD still needs to decide how to best organize, lead, and support space activities. If it does not do so, its commitment to reforms may not be sustainable.

SPACE ACQUISITION CHALLENGES PERSIST

A long-standing problem in DOD space acquisitions is that program and unit costs tend to go up significantly from initial cost estimates, while in some cases, the capability that was to be produced goes down. Figures 1 and 2 reflect differences in total program and unit costs for satellites from the time the programs officially began to their most recent cost estimates.

As figure 1 shows, in several cases, DOD has had to cut back on quantity and capability in the face of escalating costs. For example, two satellites and four instruments were deleted from the National Polar-orbiting Operational Environmental Satellite System (NPOESS) and four sensors are expected to have fewer capabilities. This will reduce some planned capabilities for NPOESS as well as planned coverage. The figures below reflect the total program costs developed in fiscal year 2009. (Last year, we also compared original cost estimates to current cost estimates for the broader portfolio of major space acquisitions for fiscal years 2008 through 2013. However, we were unable to perform this analysis this year because, for most of its major weapon system programs, DOD in fiscal year 2009 did not issue complete Selected Acquisition Reports, which contain updated yearly program funding estimates needed to conduct the analysis.)

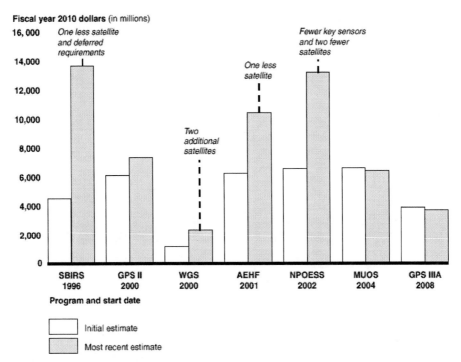

Source: GAO analysis of DOD data.

Legend: SBIRS = Space Based Infrared System; GPS = Global Positioning System;
WGS = Wideband Global SATCOM; AEHF = Advanced Extremely High
Frequency; NPOESS = National Polar-orbiting Operational Environmental
Satellite System; MUOS = Mobile User Objective System.

Figure 1. Differences in Total Program Costs from Program Start and Most Recent
Estimates (Fiscal Year 2009).

Several space acquisition programs are years behind schedule. Figure 3
highlights the additional estimated months needed for programs to deliver
initial operational capabilities (IOC). These additional months represent time
not anticipated at the programs' start dates. Generally, the further schedules
slip, the more DOD is at risk of not sustaining current capabilities. For
example, according to Air Force officials, they have requested information
from the space community on how best to address a potential gap in missile
warning capabilities.

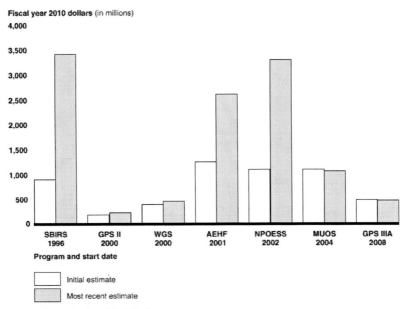

Source: GAO analysis of DOD data.

Legend: SBIRS = Space Based Infrared System; GPS = Global Positioning System; WGS = Wideband Global SATCOM; AEHF = Advanced Extremely High Frequency; NPOESS = National Polar-orbiting Operational Environmental Satellite System; MUOS = Mobile User Objective System.

Figure 2. Differences in Unit Costs from Program Start to Most Recent Estimates (Fiscal Year 2009).

Some Acquisition Programs Have Overcome Problems and Have Satellites Ready for Launch

DOD has made progress on several of its high-risk space programs and is expecting significant advances in capability as a result. In 2009, DOD launched the third Wideband Global SATCOM (WGS) satellite, broadening communications capability available to warfighters—and a fourth WGS satellite is slated for launch in 2011. DOD also launched two Global Positioning System (GPS) IIR-M satellites, although one has still not been declared operational because of radio signal transmission problems. Lastly, DOD supported the launch of a pair of Space Tracking and Surveillance System satellites, designed to test the tracking of ballistic missiles in support of missile defense early missile warning missions— these suffered many

delays as well. The Evolved Expendable Launch Vehicle (EELV) program had
its 31st consecutive successful operational launch last week.

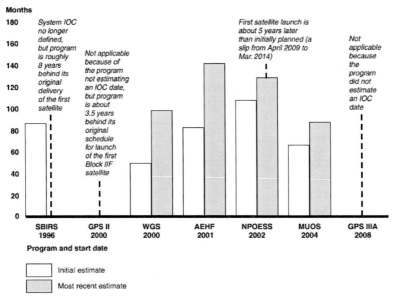

Source: GAO analysis of DOD data.
Legend: SBIRS = Space Based Infrared System; GPS = Global Positioning System;
WGS = Wideband Global SATCOM; AEHF = Advanced Extremely High
Frequency; NPOESS = National Polar-orbiting Operational Environmental
Satellite System; MUOS = Mobile User Objective System.

Figure 3. Differences in Total Number of Months to IOC from Program Start and Most
Recent Estimates.

Moreover, though it has had long-standing difficulties on nearly every
space acquisition program, DOD now finds itself in a position to possibly
launch the first new satellite from four different major space acquisition
programs over the next 12 months that are expected to significantly contribute
to missions and capabilities. These include the Global Positioning System
(GPS) IIF satellites, the Advanced Extremely High Frequency (AEHF)
communications satellites, and the Space Based Space Surveillance (SBSS)
satellite—all of which struggled for years with cost and schedule growth,
technical or design problems, as well as oversight and management
weaknesses. Table 1 further describes the status of these efforts.

**Table 1. Systems Nearing Launch That Have
Overcome Technical and Other Problems**

GPS IIF	The first GPS IIF satellite is expected to launch in mid-2010 and will upgrade timing and navigation accuracy, and add a new signal for civilian use. The satellite has been delayed over 3 years from its original launch date to May 2010—representing a further 6 month slip since we reported last year. Also, the cost of the GPS IIF program is now expected to be about $1.7 billion—almost $1 billion over the original cost estimate of $729 million. (This approximately 133 percent cost increase is not apparent in figures 1 and 2 because the GPS II modernization program includes the development and procurement of 33 satellites, only 12 of which are IIF satellites.) According to the GPS Wing, the remaining technical issues with the first IIF satellite were resolved and will not affect the scheduled launch date—the last technical issue was a desire to provide additional fault protection and this is being addressed with enhanced ground operations procedures. Additionally, the GPS Wing stated that the ground control software needed to support the first IIF launch has been thoroughly tested and in place since early this month.
AEHF	AEHF, which appears to have overcome its technical problems that delayed the first satellite's launch and increased program cost, is expected to launch in September 2010, and is expected to deliver 10 times the communications bandwidth that is available today for secure and protected communications. The launch of the first satellite has slipped almost 6 years. DOD intends to buy three more satellites, bringing the total to six (two of these additional satellites are not reflected in figures 1 and 2). The program has decided that the design specifications for the first three satellites will remain unchanged for satellites four through six, which will thus be clones except for the replacement of obsolete parts. The program office estimates that the fourth AEHF satellite will cost significantly more than the third satellite because some components that are no longer manufactured will have to be replaced and production will have to be restarted after a 4-year gap. Because of these delays, IOC has slipped about 5 years—from 2008 to 2013. The AEHF program office estimates the cost of the fifth and sixth satellites to be about $1.6 billion and $1.7 billion (then-years dollars), with estimated launch dates in 2018 and 2020, respectively.
SBSS	The first SBSS Block 10 satellite is expected to launch in 2010 and is expected to provide greatly improved space situational awareness to help better understand location and mission capabilities of all satellites and other objects in space. The launch is expected to be about 3 years later than originally planned—in part because of launch vehicle issues unrelated to the satellite. Program officials and the SBSS contractors are studying the feasibility of launching the SBSS satellite on a Delta II rocket. The program was restructured in 2006 after an independent review found that the requirements were overstated and its cost and schedule targets could not be met.

Source: GAO analysis of DOD data and previous GAO reports.

One program that appears to be overcoming remaining technical problems, but for which we are still uncertain whether it can meet its current launch date, is the Space Based Infrared System (SBIRS) satellite program. The first of four geosynchronous earth-orbiting (GEO) satellites (two sensors have already been launched on a highly elliptical orbit) is expected to launch in December 2010 and is expected to continue the missile warning mission with sensors that are more capable than the satellites currently on orbit. Total cost for the SBIRS program is currently estimated at over $13.6 billion for four GEO satellites (and two sensors that have already been delivered and are operational), representing an increase of about $9.2 billion over the program's original cost, which included five GEO satellites. The most recent program estimate developed in 2008 set December 2009 as the launch goal for the first GEO satellite, but program officials indicate that the first GEO launch will be delayed at least another year, bringing the total delay to approximately 8 years. The reasons for the delay include poor government oversight of the contractor, technical complexities, and rework. The program continues to struggle with flight software development, and during testing last year, officials discovered hardware defects on the first GEO satellite, though the program reports that they have been resolved. The launches of subsequent GEO satellites have also slipped as a result of flight software design issues. Program officials indicate that they again intend to re-baseline the program to more realistic cost and schedule estimates by mid- to late-2010. Because of the problems on SBIRS, DOD began a follow-on system effort, now known as Third Generation Infrared Surveillance (3GIRS), to run in parallel with the SBIRS program. For fiscal year 2011, DOD plans to cancel the 3GIRS effort, but also plans to provide funds under the SBIRS program for one of the 3GIRS infrared demonstrations nearing completion.

Other Programs Still Susceptible to Cost and Schedule Overruns

While DOD is having success in readying some satellites for launch, other space acquisition programs face challenges that could further increase cost and delay targeted delivery dates. The programs that may be susceptible to cost and schedule challenges include NPOESS, Mobile User Objective System (MUOS), and GPS IIIA. Delays in both the NPOESS and MUOS programs have resulted in critical potential capability gaps for military and other government users. The GPS IIIA program was planned with an eye toward avoiding problems that plagued the GPS IIF program, but the schedule leaves little room for potential problems and there is a risk that the ground system needed to operate the satellites will not be ready when the first satellite is launched. Table 2 describes the status of these efforts in more detail.

Table 2. Programs Still Susceptible to Cost and Schedule Overruns

NPOESS	The NPOESS program has continued to experience technical problems resulting in further cost and schedule increases. The program was restructured in 2007, which led to a reduction in the number of satellites from six to four and deletions or replacements of satellite sensors. NPOESS was originally estimated to cost $6.5 billion but the latest estimate is about $13.2 billion—representing more than a 100-percent cost increase. Furthermore, the launch of the first satellite has slipped about 5 years—from April 2009 to March 2014. While the goal of the restructure was to lower future cost and schedule risks, it increased the risk of a satellite coverage gap and significantly reduced data collection capabilities. DOD programmed funds for NPOESS for fiscal year 2011, but according to the White House's Office of Science and Technology Policy, the NPOESS program is to be restructured. This would allow DOD and the Department of Commerce to embark on separate weather satellite programs to meet their unique needs. The cost and schedule estimates for the NPOESS program cited above do not reflect the latest events surrounding the program. At this juncture, many questions surround DOD's strategy for moving forward, including the following: (1) How does DOD intend to use the funding programmed for fiscal year 2011? (2) Is the NPOESS contract to be terminated, and if so, what are the anticipated termination costs for work under contract? (3) What aspects of the NPOESS program will continue to be utilized for future efforts? (4) Will the approach going forward be more or less costly, and will the delivery of capability be sooner or later than that of NPOESS? While many of these details have yet to be worked out, this major redirection so late in the acquisition process may pose significant risk to the nation's ability to reconstitute its weather satellites in a timely fashion to mitigate lapses in data collection capabilities.
MUOS	The MUOS communications satellite program now estimates a 21-month delay— from March 2010 to December 2011—in the delivery of on-orbit capability from the first satellite. This represents an additional 10-month slip from the slip we reported last year, which was caused by continuing satellite development challenges. In July 2009, a Navy-initiated review of the program found that while the technical challenges the program was experiencing could be solved, the MUOS budget was inadequate and its schedule was optimistic. Subsequently, in late 2009 the Navy established new cost and schedule baselines for the program (we have yet to obtain the new cost baseline, and as such, figures 1 and 2 do not reflect updated MUOS cost estimates). In January 2011, communications are predicted to degrade below the required level of availability and remain so until the first MUOS satellite is available for operations. The MUOS program office is addressing the potential capability gap by activating dual digital receiver unit operations on a legacy satellite, leasing commercial ultra-high-frequency satellite communications services, and examining the feasibility of expanded digital receiver unit operations on the legacy payloads of the MUOS satellites.

GPS IIIA	While the GPS IIIA program has been structured by the Air Force to prevent the mistakes made on the IIF program, the Air Force aims to deliver the GPS IIIA satellites 3 years faster than the IIF satellites. According to Air Force officials, the IIIA contractor retained some of its workforce from the IIR-M program and plans to incorporate a previously developed satellite bus—efforts that reduce program risk. However, we continue to believe the IIIA schedule is optimistic given the program's late start, past trends in space acquisitions, and challenges facing the new contractor. To increase confidence in the schedule for delivering the ground control system for IIIA (the next generation operational control segment known as OCX), the GPS Wing added 16 months of development time to the effort. This means that OCX is now scheduled to be fielded after the May 2014 launch of the first GPS IIIA satellite. The Wing is currently assessing alternate approaches for resolving the fielding issue, which will likely have cost consequences.

Source: GAO analysis of DOD data and previous GAO reports.

a GAO, Global Positioning System: Significant Challenges in Sustaining and Upgrading Widely Used Capabilities, GA0-09-325, (Washington, D.C.: April 30, 2009).

Challenges in Aligning Space System Components

This past year we also assessed the levels at which DOD's satellites, ground control, and user terminals were synchronized to provide maximum benefit to the warfighter.[1] Most space systems consist of satellites, ground control systems, and user terminals, though some space systems only require ground control systems to provide capability to users. Ground control systems are generally used to (1) download and process data from satellite sensors and disseminate this information to warfighters and other users and (2) maintain the health and status of the satellites, including steering the satellites and ensuring that they stay in assigned orbits.

User terminals, typically procured by the military services and managed separately from associated satellites and ground control systems, can range from equipment hosted on backpacks to terminals mounted on Humvees, airborne assets, or ships. Terminals can be used to help the warfighter determine longitude, latitude, and altitude via GPS satellites, or securely communicate with others via AEHF satellites. Some user terminals are not solely dedicated to delivering capability from a specific satellite system. For example, the Joint Tactical Radio System is the primary user terminal associated with the MUOS program, but the system is also designed to be the next generation of tactical radios, allowing extensive ground-to-ground communication as well.

Overall, we found the alignment of space system components proved to be challenging to DOD. Specifically, we found that for six of DOD's eight major space system acquisitions, DOD has not been able to align delivery of satellites with ground control systems, user terminals, or both. Of the eight major space system acquisitions, five systems' ground control system efforts are optimally aligned to deliver capability with their companion satellites, while three are not. For the five space systems requiring user terminals, none was aligned. In some cases, capability gaps of 4 or more years have resulted from delays in the fielding of ground control systems or user terminals. When space system acquisitions are not aligned, satellite capability is available but underutilized, though in some cases, workaround efforts can help compensate for the loss or delay of capability. Moreover, when ground systems, user terminals, or both are not aligned with satellites, there are significant limitations in the extent to which the system as a whole can be independently tested and verified.[2,3]

Launch Manifest Issues

Another risk facing DOD space programs for the next few years is the potential for increased demand for certain launch vehicles. DOD is positioned to launch a handful of satellites across missions over the next 2 years that were originally scheduled for launch years ago. Until recently, DOD had four launch pads on the East Coast from which to launch military satellites. In 2009, DOD launched the final two GPS IIR-M satellites using the Delta II launch vehicle, thereby discontinuing its use of the Delta II line and its associated launch infrastructure. DOD now plans to launch most of its remaining satellites using one of DOD's EELV types— Atlas V or Delta IV— from one of two East Coast launch pads. At the same time, the National Aeronautics and Space Administration (NASA) plans to use the Delta II to launch at least three major missions before that launch vehicle is retired. In addition, NASA is already manifesting other major missions on the Atlas V. Given the expected increased demand for launches—many of which are considered high priority—and the tempo of launches DOD has achieved with EELV, it appears that the launch manifest is crowded. As a result, if programs still struggling with technical, design, or production issues miss their launch dates, the consequences could be significant, as it may take many months to secure new dates. Some of DOD's satellites are dual integrated, which means they can be launched on either type of EELV. The Air Force deserves credit for designing the satellites this way because it offers more flexibility in terms of launch vehicle usage, but there are also cost and schedule implications associated with rescheduling from one EELV type to the other. Moreover, DOD can request its launch provider to speed up the transition time between launches, although this would also increase costs. Nevertheless, Air Force officials stated that they were confident that the higher launch rates could be achieved, especially if a particular satellite's priority increased. According to Air Force officials, they have already begun to implement means to address these issues.

DOD Is Taking Actions to Address Space and Weapon Acquisition Problems

DOD has been working to ensure that its space programs are more executable and produce a better return on investment. Many of the actions it is taking address root causes of problems, though it will take time to determine whether these actions are successful and they need to be complemented by decisions on how best to lead, organize, and support space activities.

Our past work has identified a number of causes behind the cost growth and related problems, but several consistently stand out. First, on a broad scale, DOD starts more weapon programs than it can afford, creating a competition for funding that encourages low cost estimating, optimistic scheduling, overpromising, suppressing bad news, and for space programs, forsaking the opportunity to identify and assess potentially more executable alternatives. Second, DOD has tended to start its space programs too early, that is, before it has the assurance that the capabilities it is pursuing can be achieved within available resources and time constraints. This tendency is caused largely by the funding process, since acquisition programs attract more dollars than efforts concentrating solely on proving technologies. Nevertheless, when DOD chooses to extend technology invention into acquisition, programs experience technical problems that require large amounts of time and money to fix. Moreover, there is no way to accurately estimate how long it would take to design, develop, and build a satellite system when critical technologies planned for that system are still in relatively early stages of discovery and invention. Third, programs have historically attempted to satisfy all requirements in a single step, regardless of the design challenge or the maturity of the technologies necessary to achieve the full capability. DOD has preferred to make fewer but heavier, larger, and more complex satellites that perform a multitude of missions rather than larger constellations of smaller, less complex satellites that gradually increase in sophistication. This has stretched technology challenges beyond current capabilities in some cases and vastly increased the complexities related to software. Programs also seek to maximize capability on individual satellites because it is expensive to launch.

In addition, problematic implementation of an acquisition strategy in the 1990s, known as Total System Performance Responsibility, for space systems resulted in problems on a number of programs because it was implemented in a manner that enabled requirements creep and poor contractor performance— the effects of which space programs are still addressing. We have also reported on shortfalls in resources for testing new technologies, which coupled with less expertise and fewer contractors available to lead development efforts, have magnified the challenge of developing complex and intricate space systems.

Our work—which is largely based on best practices in the commercial sector—has recommended numerous actions that can be taken to address the problems we identified. Generally, we have recommended that DOD separate technology discovery from acquisition, follow an incremental path toward meeting user needs, match resources and requirements at program start, and use quantifiable data and demonstrable knowledge to make decisions to move to next phases. We have also identified practices related to cost estimating, program manager tenure, quality assurance, technology transition, and an array of other aspects of acquisition program management that could benefit space programs. These practices are detailed in appendix I.

DOD is implementing an array of actions to reform how weapons and space systems are acquired. For space in particular, DOD is working to ensure critical technologies are matured before large-scale acquisition programs begin; requirements are defined early in the process and are stable throughout; and that system design remains stable, according to the Director of Space and Intelligence under DOD's Office of the Secretary of Defense for Acquisition, Technology and Logistics. DOD also intends to follow incremental or evolutionary acquisition processes versus pursuing significant leaps in capabilities involving technology risk. The Director of Space and Intelligence also told us that DOD is revisiting the use of military standards in its acquisitions and providing more program and contractor oversight. The approach described to us by the Director of Space and Intelligence mirrors best practices identified in our reports. Moreover, some actions—described in the table below—have already been taken to ensure acquisitions are more knowledge-based.

Table 3. Actions being Taken to Address Space Acquisition Problems

Requirements	The Air Force leadership signed the Acquisition Improvement Plan which lists five initiatives for improving how the Air Force obtains new capabilities—one of these initiatives covers requirements generation and includes the direction for the Air Force to certify the acquisition community can successfully fulfill required capabilities in conjunction with the Air Force Requirements for Operational Capabilities Council. Certification means the required capabilities can be translated in a clear and unambiguous way for evaluation in a source selection, are prioritized if appropriate, and organized into feasible increments of capability.
Program Management	The Space and Missile Systems Center—the Air Force's primary organization responsible for acquiring space systems—resurrected a program management assistance group in 2007 to help mitigate program management, system integration, and program control deficiencies within specific ongoing programs. This group assists and supplements wing commanders and program offices in fixing common problems, raising core competencies, and providing a consistent culture that sweeps across programs. According to the GPS Wing Commander, this group was an integral part of the overall process providing applicationoriented training, templates, analyses, and assessments vital to the GPS IIIA baseline review.
Workforce	The Air Force is continuing efforts to bring space operators and space system acquirers together through the Advanced Space Operations School and the National Security Space Institute. The Air Force anticipates that this higher-level education will be integral to preparing space leaders with the best acquisition know-how.
Cost Estimating	Both the Air Force and the National Reconnaissance Office (NRO) are taking actions to strengthen costestimating. For example, we recommended that the Secretary of the Air Force ensure that cost estimates are updated as major events occur within a program that could have a material impact on cost, and that the roles and responsibilities of the various Air Force cost-estimating organizations be clearly articulated.[a] An Air Force policy directive now requires that cost estimates for major programs be updated annually, and lays out roles and responsibilities for Air Force cost-estimating organizations. Furthermore, in its attempts to make more accurate cost estimates for commercial-like programs (characterized by use of fixed-price contracts, less complex satellites, lower costs, and short development timeframes), the NRO cost analysis improvement group has developed a cost-estimating methodology that considers acquisition complexity (such as level of oversight and amount of program reporting), in addition to program technical complexity, and stated it is considering applying the methodology to more traditional satellite acquisition programs.

Table 3. (Continued)

Acquisition Policy	DOD recently eliminated the tailored national security space acquisition policy and moved the acquisition of space systems under DOD's updated acquisition guidance for defense acquisition programs (DOD Instruction 5000.02). DOD is currently developing an addendum for the Instruction that would introduce specific management and oversight processes for acquiring major space systems.
Alignment of Ground Control Systems	In better aligning space system components, DOD acknowledged that the integration and consolidation of satellite ground control systems has many benefits, and established the Space and Intelligence Office to more effectively conduct oversight of the space and intelligence enterprise. DOD further disestablished two oversight boards that were deemed less effective in providing oversight.

Source: GAO analysis of DOD data and previous GAO reports.

[a] GAO, Space Acquisitions: DOD Needs to Take More Action to Address Unrealistic Initial Cost Estimates of Space Systems, GAO-07-96, (Washington, D.C.: November 17, 2006).

Congress has also acted on a broader scale through the Weapon Systems Acquisition Reform Act, which was signed into law on May 22, 2009.[4] The goal of this new statute is to improve acquisition outcomes in DOD, with specific emphasis on major defense acquisition programs (MDAP) and major automated information systems. According to the President of the United States this legislation is designed to limit cost overruns before they spiral out of control and will strengthen oversight and accountability by appointing officials who will be charged with closely monitoring the weapons systems being purchased to ensure that costs are controlled. DOD states in its 2010 Quadrennial Defense Review[5] that the law also will substantially improve the oversight of major weapons acquisition programs, while helping to put MDAPs on a sound footing from the outset by addressing program shortcomings in the early phases of the acquisition process. DOD also states that it is undertaking a far-reaching set of reforms to achieve these goals and to improve how DOD acquires and fields critical capabilities for current and future wars and conflicts.

ADDITIONAL DECISIONS ON LEADERSHIP, ORGANIZATION, AND SUPPORT ARE STILL NEEDED

The actions that the Air Force and Office of the Secretary of Defense have been taking to address acquisition problems are good steps. However, there are still more significant changes to processes, policies, and support needed to ensure that reforms can take hold. Recent studies and reviews examining the leadership, organization, and management of national security space have all found that there is no single authority responsible below the President and that authorities and responsibilities are spread across the department. In fact, the national security space enterprise comprises a wide range of government and nongovernment organizations responsible for providing and operating space-based capabilities serving both military and intelligence needs.

In 2008, for example, a congressionally chartered commission (known as the Allard Commission)[6] reported that responsibilities for military space and intelligence programs were scattered across the staffs of DOD organizations and the intelligence community and that it appeared that "no one is in charge" of national security space. The same year, the House Permanent Select Committee on Intelligence reported similar concerns, focusing specifically on difficulties in bringing together decisions that would involve both the Director of National Intelligence and the Secretary of Defense.[7] Prior studies, including those conducted by the Defense Science Board and the Commission to Assess United States National Security Space Management and Organization (Space Commission),[8] have identified similar problems, both for space as a whole and for specific programs. While these studies have made recommendations for strengthening leadership for space acquisitions, no major changes to the leadership structure have been made in recent years. In fact, an executive agent position within the Air Force that was designated in 2001 in response to a Space Commission recommendation to provide leadership has not been filled since the last executive resigned in 2007.

Diffuse leadership has a direct impact on the space acquisition process, primarily because it makes it difficult to hold any one person or organization accountable for balancing needs against wants, for resolving conflicts among the many organizations involved with space, and for ensuring that resources are dedicated where they need to be dedicated. Many of the cost and schedule problems we identified for the GPS IIF program, for instance, were tied in part to diffuse leadership and organizational stovepipes, particularly with respect to DOD's ability to coordinate delivery of space, ground, and user assets. In fact,

DOD is now facing a situation where satellites with advances in capability will be residing for years in space without users being able to take full advantage of them because investments and planning for ground, user, and space components were not well-coordinated.

Congressional and DOD studies have also called for changes in the national security space organizational structure to remove cultural barriers to coordinating development efforts and to better incorporate analytical and technical support from an organization that is augmented with military and intelligence community expertise.

Finally, studies have identified insufficient numbers of experienced space acquisition personnel and inadequate continuity of personnel in project management positions as problems needing to be addressed in the space community. Our own studies have identified gaps in key technical positions, which we believed increased acquisition risks. For instance, in a 2008 review of the EELV program, we found that personnel shortages at the EELV program office occurred particularly in highly specialized areas, such as avionics and launch vehicle groups.[9] These engineers work on issues such as reviewing components responsible for navigation and control of the rocket. Moreover, only half the government jobs in some key areas were projected to be filled. These and other shortages in the EELV program office heightened concerns about DOD's ability to effectively manage the program using a contracting strategy for EELV that required greater government attention to the contractor's technical, cost, and schedule performance information. In a recent discussion with GAO, the Director of Space and Intelligence under DOD's Office of the Secretary of Defense for Acquisition, Technology and Logistics stated that the primary obstacle to implementing reforms in space is the lack of "bench strength," primarily technical and systems engineering expertise.

CONCLUDING REMARKS

In conclusion, DOD space is at a critical juncture. After more than a decade of acquisition difficulties, which have created potential gaps in capability, diminished DOD's ability to invest in new space systems, and lessened DOD's credibility to deliver high-performing systems within budget and on time, DOD is finally positioned to launch new generations of satellites that promise vast enhancements in capability. Moreover, recent program cancellations have alleviated competition for funding and may have allowed

DOD to focus on fixing problems and implementing reforms rather than taking on new, complex, and potentially higher-risk efforts. But these changes raise new questions. Specifically, when can investments in new programs be made? How can reforms really take hold when leadership is diffuse? How can reforms take hold when there are still organizational barriers that prevent effective coordination? And lastly, how can acquisitions be successful if the right technical and programmatic expertise is not in place? Clearly, there are many challenges ahead for space. We look forward to working with the DOD to help ensure that these and other questions are addressed.

Mr. Chairman, this concludes my prepared statement. I would be happy to answer any questions you or members of the subcommittee may have at this time.

APPENDIX I: ACTIONS NEEDED TO ADDRESS SPACE AND WEAPON ACQUISITION PROBLEMS

Before undertaking new programs
• Prioritize investments so that projects can be fully funded and it is clear where projects stand in relation to the overall portfolio.
• Follow an evolutionary path toward meeting mission needs rather than attempting to satisfy all needs in a single step.
• Match requirements to resources—that is, time, money, technology, and people—before undertaking a new development effort.
• Research and define requirements before programs are started and limit changes after they are started.
• Ensure that cost estimates are complete, accurate, and updated regularly.
• Commit to fully fund projects before they begin.
• Ensure that critical technologies are proven to work as intended before programs are started.
• Assign more ambitious technology development efforts to research departments until they are ready to be added to future generations (increments) of a product.
• Use systems engineering to close gaps between resources and requirements before launching the development process.
During program development
• Use quantifiable data and demonstrable knowledge to make go/no-go decisions, covering critical facets of the program such as cost, schedule, technology readiness, design readiness, production readiness, and relationships with suppliers.
• Do not allow development to proceed until certain thresholds are met—for example, a high proportion of engineering drawings completed or production processes under statistical control.

Appendix I. (Continued)

• Empower program managers to make decisions on the direction of the program and to resolve problems and implement solutions.
• Hold program managers accountable for their choices.
• Require program managers to stay with a project to its end.
• Hold suppliers accountable to deliver high-quality parts for their products through such activities as regular supplier audits and performance evaluations of quality and delivery, among other things.
• Encourage program managers to share bad news, and encourage collaboration and communication.

Source: GAO.

APPENDIX II: SCOPE AND METHODOLOGY

In preparing this testimony, we relied on our body of work in space programs, including previously issued GAO reports on assessments of individual space programs, common problems affecting space system acquisitions, and the Department of Defense's (DOD) acquisition policies. We relied on our best practices studies, which comment on the persistent problems affecting space acquisitions, the actions DOD has been taking to address these problems, and what remains to be done, as well as Air Force documents addressing these problems and actions. We also relied on work performed in support of our annual weapons system assessments, and analyzed DOD funding estimates to assess cost increases and investment trends for selected major space acquisition programs. The GAO work used in preparing this statement was conducted in accordance with generally accepted government auditing standards. Those standards require that we plan and perform the audit to obtain sufficient, appropriate evidence to provide a reasonable basis for our findings and conclusions based on our audit objectives. We believe that the evidence obtained provides a reasonable basis for our findings and conclusions based on our audit objectives.

End Notes

[1] GAO, Defense Acquisitions: Challenges in Aligning Space System Components, GAO-10-55 (Washington, D.C.: Oct. 29, 2009).

[2] In making determinations about whether space system acquisitions were aligned, we examined whether there were gaps between fielding dates of satellite capabilities compared to ground system capabilities and whether lower percentages of user terminal types were planned to

be fielded by the space system acquisitions' planned initial capability. Generally we considered aspects of a space acquisition unaligned if there was a gap of years, rather than months, between the fielding dates of significant capabilities. Regarding user terminals, we only considered these unaligned compared to satellite capabilities when user terminals did not meet DOD's measure of synchronization for military satellite communications space acquisitions. This measure, established by the U.S. Strategic Command, a primary user of DOD space systems, asserts that 20 percent of any type of user terminal should be fielded by a space system acquisition's initial capability date and 85 percent should be fielded by its full capability date.

[3] It should be noted that while there are criteria for communications satellites, there are no criteria available in DOD that determine the optimum alignment or synchronization for the broader portfolio of satellite programs. This is principally because of inherent differences in satellite missions and their associated ground and user assets, according to officials involved in space system development as well as acquisition oversight.

[4] Pub. L. No. 111-23, 123 Stat. 1704 (2009).

[5] Department of Defense, Quadrennial Defense Review Report (Washington, D.C., Feb. 1, 2010).

[6] Institute for Defense Analyses, Leadership, Management, and Organization for National Security Space: Report to Congress of the Independent Assessment Panel on the Organization and Management of National Security Space (Alexandria, VA., July 2008).

[7] House Permanent Select Committee on Intelligence, Report on Challenges and Recommendations for United States Overhead Architecture (Washington, D.C., October 2008).

[8] Department of Defense, Report of the Commission to Assess United States National Security Space Management and Organization (Washington, D.C., Jan. 11, 2001).

[9] GAO, Space Acquisitions: Uncertainties in the Evolved Expendable Launch Vehicle Program Pose Management and Oversight Challenges, GAO-08-1039 (Washington, D.C.: Sept. 26, 2008).

INDEX